The Energy Transition

Topical Conference at the 2023 AIChE Spring Meeting and 19th Global Congress on Process Safety

Houston, Texas, USA
12-16 March 2023

ISBN: 978-1-7138-7938-1

Printed from e-media with permission by:

Curran Associates, Inc.
57 Morehouse Lane
Red Hook, NY 12571

Some format issues inherent in the e-media version may also appear in this print version.

Printed with permission by Curran Associates, Inc. (2023)

For permission requests, please contact AIChE
at the address below.

AIChE
120 Wall Street, FL 23
New York, NY 10005-4020

Phone: (800) 242-4363
Fax: (203) 775-5177

www.aiche.org

Additional copies of this publication are available from:

Curran Associates, Inc.
57 Morehouse Lane
Red Hook, NY 12571 USA
Phone: 845-758-0400
Fax: 845-758-2633
Email: curran@proceedings.com
Web: www.proceedings.com

TABLE OF CONTENTS

NEW TECHNOLOGIES IN ENERGY TRANSITION

FEEDSTOCK EXPANSION FOR ENERGY TRANSITION

19a Invited: Energy Transition Roundtable with Energy Leaders

Joseph Powell, Ignasi Palou-Rivera, Faruque Hasan, John D. Rogers

This session is setup for a roundtable introductory discussion on Energy Transition and present perspectives from the current industry, academic and governmental leaders. These will be in the format of 10 minute presentations to set the stage for discussions. Then, the entire audience is invited to participate in six breakout sessions that will discuss the following:

Feedstock Expansion for Energy Transition
New Technologies in Energy Transition
Transportation considerations in Energy Transition
Design for Energy Transition
Modeling for Energy Transition
Analysis for Energy Transition

We hope to hear from leaders in the energy, fuels, and petrochemicals sector, engineers, researchers, and government labs to create a path forward for the above topics of interest. Please join us in providing your inputs and perspectives, that will be shared as a workshop proceeding with the AIChE F&PD and other organization wide initiatives on Energy Transition.

Invited Panelist Bio:

Dr. Joe Powell (Joseph B. Powell, PhD)

Energy Transition Institute, University of Houston

National Academy of Engineering, Fellow AIChE

Retired Shell Chief Scientist – Chemical Engineering JoePowell-Linkedin; JBPowel5@Central.UH.edu

Joe Powell (Joseph B. Powell, PhD) is Executive Director of the University of Houston Energy Transition Institute, a member of the U.S. National Academy of Engineering, Fellow and former Director of the American Institute of Chemical Engineers. He served as Shell's first Chief Scientist – Chemical Engineering from 2006 – 2020, culminating a 36-year industry career where he led R&D programs in new chemical processes, biofuels, enhanced oil recovery, and advised on global strategy for the energy transition to a net-zero carbon economy. He is co-inventor on more than 125 patent applications (60 granted), has received AIChE / ACS / R&D Magazine awards for Innovation, Service, and Practice, and is co-author of Sustainable Development in the Process Industries: Cases and Impact (2010). He chaired the U.S. Department of Energy Hydrogen and Fuel Cell Technical Advisory Committee (HTAC), served two terms on the U. S. National Academy Board on Chemical Sciences and Technology and on the editorial board of Annual Review of Chemical and Biological Engineering, and serves as climate advisor for the U.S. Business Council for Sustainable Development. He served as crosscutting team lead for Mission Innovation Carbon Capture Utilization and Storage (2017), and currently serves on the National Academy Carbon Utilization Infrastructure, Markets, Research, and Development Committee. Dr. Powell obtained a PhD from the U. Wisconsin-Madison (1984); and a BS from the U. Virginia (1978), both in chemical engineering.

Dr. Faruque Hasan

Assistant Director of Decarbonization, Texas A&M Energy Institute

Associate Professor, Kim and Phillip McDivitt Faculty Fellow, Artie McFerrin Department of Chemical Engineering, Texas A&M University

Dr. Faruque Hasan is an Assistant Director of Decarbonization at the Texas A&M Energy Institute, and an Associate Professor and the Kim and Phillip McDivitt Faculty Fellow in the Artie McFerrin Department of Chemical Engineering at Texas A&M University. He holds a Ph.D. in Chemical Engineering from National University of Singapore, and completed a postdoctoral training at Princeton University before joining Texas A&M University in 2014. At Texas A&M, he directs a research group that is recognized for developing multiscale methods, tools and techniques for systems design and analysis with direct applications to energy transition and decarbonization pathways for the chemical process industry, carbon capture, utilization and storage (CCUS), sustainable hydrogen economy, and resilient supply chains. He is a Senior Member and also serves as a Director for the Computing & Systems Technology Division of the American Institute of Chemical Engineers (AIChE). He is the recipient of NSF CAREER award, and AIChE Computing and Systems Technology (CAST) Division Outstanding Young Research award.

Dr. Ignasi Palou-Rivera

Executive Director and Chief Technology Officer, RAPID Manufacturing Institute

Ignasi Palou-Rivera is the Executive Director and Chief Technology Officer of the RAPID Manufacturing Institute. Ignasi's professional career of over 25 years spans the areas of process development, process modeling and optimization, refinery planning and scheduling, technology valuation and commercialization covering techno-economic, life-cycle, sustainability analysis, and tech to market activities, as well as R&D management. He has worked for a variety of organizations such as Aspentech, Applied Materials, BP, LanzaTech and Argonne National Laboratory prior to joining RAPID in 2017. Ignasi is a Senior Member of AIChE and has been heavily involved as a volunteer both in the Computing & Systems Technology Division (CAST) and the Sustainable Engineering Forum (SEF) AIChE Communities, especially in programing functions. Ignasi is a Past Chair of the Sustainable Engineering Forum, part of the Institute for Sustainability (IfS) Board, and member of the PAIC Policy Committee. Ignasi holds an Engineer degree from the Universitat Politècnica de Catalunya in Barcelona and a Ph.D. in Chemical Engineering from the University of Wisconsin-Madison and was a post-doctoral researcher at MIT.

Dr. John D. Rogers

Technology Manager, Science & Technology Strategic Plans & Programs Division, US DOE National Energy Technology Laboratory

John D. Rogers is a Technology Manager in the Science & Technology Strategic Plans & Programs Division at the US DOE National Energy Technology Laboratory in the Houston, Texas, office. Currently his responsibility is the Advanced Remediation Technologies portfolio of programs. He is a Professional Engineer (PE) and holds a Ph.D. and B.S. in Chemical Engineering from New Mexico State University, as well as an M.S. in Petroleum Engineering from Texas Tech University in Lubbock. His career has developed diverse roles in federal and local government, university R&D, a non-profit, and the industrial private sector where he primarily focused on environmentally responsible resource recovery. Rogers has held positions as a Senior Production Operations Engineer, Reservoir Engineer, Industrial Wastewater Control Supervisor, Project/Program manager, and Vice President of Operations. With his commitment to sustainable clean energy, he is dedicated to fostering team success in creating and utilizing cutting edge technology to develop technically sustainable energy resources.

In his current role, he has the opportunity to lead a cross functional interdisciplinary technical team within NETL to execute the U.S. DOE's FECM Advanced Remediation Technologies research direction. As Technology Manager, he oversees the strategic and technical direction and implementation of the R&D in the EPD and Hydrates programs at NETL and ensures government funding is obligated in a timely and fiscally responsible manner. The current focus of the EPD program is coordinating research and advising technical direction to maximize the benefits and minimize the environmental impacts of U.S. unconventional natural gas liquids production. A current concentrated research area is advanced technologies related to reduced water usage and efficient produced water management as well as increased resource recovery efficiency and production operations. The Hydrates program research is focused on the technical understanding of naturally occurring hydrates in Alaska and in moderately deep areas of the Gulf of Mexico.

120a Energy Savings for Steam Cracking Furnaces By Gas Turbine Integration

Vladimir Paunovic

Ethylene plants can benefit in numerous ways by integrating gas turbines (GT) with cracking furnaces. This synergy improves the power efficiency of the GT due to the furnace acting as a heat recovery unit for the exhaust gas and at the same time, the fuel demand of the furnace is substantially reduced as it is supplied with a large amount of heat and oxygen. These advantages extend to the whole plant. The plant can reduce or fully eliminate its dependence on electricity from the grid. This can be a major factor if imported power is expensive and/or unreliable. Additionally, depending on the source of imported electricity, gas turbine integration can be a way of lowering the cracker's CO_2 footprint and GHG emissions.

Technip Energies recently executed an EPC cracking furnace - gas turbine integration project in Asia. This paper discusses the advantages resulting from this project, including savings in energy, reduced emissions and the impact on NOx. It also provides a good insight into how the integration influences cracking furnace design and operation.

120b Novel Membrane System Designs to Enable Energy and Capital Efficient Carbon Capture

Christine Parrish, Josh Watson, Ken Loprete

With 40 billion tons of anthropogenic green house gases (GHG) being added to the atmosphere every year, it is vital to make existing processes carbon neutral to prevent catastrophic climate change. Hard to abate industries like cement, steel, limestone, and glass making, need a carbon dioxide solution to continue to produce the chemical building blocks for today's materials. Carbon capture technology allows users to filter and sequester the CO_2 at the source and reduce emissions by up to 90%. Hard to abate industrial sectors are unlikely to be served by electrification but are equally as unlikely to be replaced by novel technology or served by existing carbon capture technology. Compact Membrane Systems (CMS) has developed a novel Optiperm™ membrane platform to enable the efficient separation of two of the most critical separations of our industry: olefin paraffin separation and carbon capture.

Given the nature of flue gas streams, CMS chose to develop a membrane with active transport mechanism to ensure the membrane's performance at low pressure, elevated temperature, and in the presence of contaminants. This gives more flux and footprint than alternate membrane solutions and delivers practical, operational systems for carbon capture applications for less than <$50/ton. This talk will cover the performance of the membrane within the planned operating window including temperature, pressure, CO2 concentration, and stage cut. Special attention will be given to the behavior of the membrane at elevated temperature (>50°C). Initial stability data with SOx, NOx, and CO will also be examined.

While the membrane itself provides the technical unlock for viability in post combustion flue gas conditions, the system design ensures economic operation. CMS has developed a model to iterate through potential designs and minimize for a given parameter like CAPEX, annual cost of capture or electricity use. This talk will explore the sensitivities of the optimal membrane design based on CO2 feed concentration and electricity price for hard to abate emissions like cement and steel flue gas.

120c On the Design of CCS Facilities: The Case of Injection Wells, a Flow Assurance Perspective

Eduardo Luna-Ortiz, Kamila Szklarczyk-Marshall

The design of CO_2 well injectors pose particular challenges, most of them related to the thermo-physical behavior of CO_2 mixture. Such behavior in conjunction with the reservoir conditions (which varies from early injection to late life field), wellhead injection conditions and composition of the stream leads to diverse operating scenarios that are unique in Carbon Capture and Storage (CCS) systems.

In particular, in depleted reservoirs, low temperature management (in steady-state operations) due to Joule-Thomson cooling is challenging but it is, if adequately identified, it is possible to implement mitigating strategies. Typical low temperature management strategies includes wellhead heating, flowrate restriction, gas phase operation, etc. However, transient operations can also lead to conditions that often are more extreme albeit of shorter duration.

In this work, we present a number of safety-critical transient operating scenarios of the injection wells, perhaps extreme but plausible. These scenarios can pose a significant risk to the integrity of the well design. The focus of the flow assurance study was to determine the minimum fluid and metal temperatures that can occur in the well tubing during transient scenarios such as well clean-up, depressurisation, or blowout and assess the likelihood of transgressing the minimum design temperature of the well and annulus.

This first-of-a-kind CCS project is located in central Australia and it is currently in the EPC phase.

135b Digitalization: A Key Enabler for Energy Transition and Sustainability

Ravindra Aglave, John Nixon

Sustainability is an integral part of any business today. It is important for us to enable others to drive sustainable growth and transform their industries.

Our DEGREE framework is a set of strategic ambitions for the benefit of our stakeholders on the topics of environment, social and governance. As a key business that enables advanced engineering simulation, we see it playing a critical role in all aspects of engineering. We consider moving away from the two-way balance of Cost vs (Energy) Efficiency to a three-way balance of Cost-Efficiency-Sustainability a shift in the paradigm of how engineering simulation and organization wide digitalization tools can be used and applied in future of the "production-conversion-utilization" cycle in the energy and chemicals field.

In this talk, we will present first a dive into the meaning of energy transition& sustainability. We will follow that with goals and achievements in sustainability; especially highlighting how technologies in simulation & digitalization enable achievement of those goals with some examples from energy and chemical industries. And we will conclude with some thought provoking questions and challenges for young engineers.

135d A techno-economic analysis of energy efficiency and CO_2 emissions in bioethanol production due to fouling

Emilio Diaz Bejarano, Jose Loyola-Fuentes, Mariano Martin, Francesco Coletti, Carlos Prieto

Maximizing the energy efficiency and minimising GHG emissions of production processes is paramount for biofuels to become a viable economic and environmentally friendly alternative to fossil fuels, particularly in the transportation sector.

One of the main performance issues faced by these processes is the deposition of unwanted material in process equipment (fouling), especially in the beer column, evaporators and heat exchangers. This is a well know problem but its impact not well quantified in the literature. Periodic cleaning and maintenance are required to minimise shutdown time, avoid off-spec products and maintain energy consumption and CO2 emissions to levels that make the whole process economically viable (Wisconsin, 2020). However, maintenance is typically performed based on experience and simple heuristics with the main goal of ensuring production. The question arising is whether further optimisation of such maintenance activities could help further improve the average performance of production plants (by minimising water use, energy consumption, and CO2 emissions). This could be relevant to contribute to the sustainability of biofuels production and, together with other improvements, enable a true transition from conventional fuels to biofuels. However, the information available on the subject is rather dispersed (and scarce) and it is difficult to find quantitative information on the impact that fouling actually has.

In this work, after presenting information gathered in the existing literature, we aim to quantify the impact of fouling for key units in the dry-grind process based on simulations for a typical plant. The assessment indicates that fouling can easily cost in the order of several millions of dollars per year in an average size bioethanol production plant as a result of increased fuel and water consumption and extra CO2 emissions. In this context, advanced monitoring and predictive maintenance tools, which have been proven to provide significant savings to other industries, are proposed as a part of the solution to improve the management of fouling and optimise maintenance decisions.

Practical Qualitative and Quantitative Models in CCUS Risk Assessment
2023 AIChE Spring Meeting
Andy Lidstone – Risktec Solutions Ltd

Executive Summary
This presentation is intended to share some approaches that have been developed in supporting the risk management of several recent CCUS projects; in particular in adapting existing methods to apply to this emerging area.

Risk Management Concepts – As with all other risks, the risk management approach needs to be structured and proportionate. The ISO 27914 standard presents a structure which aligns with that in other similar standards such as ISO31000, including the key steps of Identification and Analysis to allow for Evaluation. Also shown is a typical proportionate approach, allowing for a range of risk based techniques to inform our decision making process.

Outline Approach – An outline of a typical approach is shown, meeting these key steps of Identification, Analysis and Evaluation, in particular with structured techniques (e.g. bowtie analysis and quantitative assessments) being used to allow for more detailed understanding of selected, key, risks.

Risk Identification – Different projects, at different phases will have the requirement to consider different risks e.g. containment, injectivity, capacity? Need to make sure that a range of potential scenarios are considered and this can be performed by structured brainstorming against a checklist, such as contained within EU Guidance Document 1, or the Features, Events and Processes database

Risk Ranking – for each scenario, there is a need to assess the level of risk, however CCS, as a relatively new field, has less operating experience than oil and gas production. A Risk Assessment Matrix allows for an initial screening, but needs to balance complexity against accuracy. The ISO 27914 standard requires as assessment of uncertainty, which can be performed at this step, but may also be included within the analysis steps.

Example Risk Register Extract – the columns are based on the guidance in ISO27914, but would need to be customised for each project's needs e.g. inherent and residual risk? Inclusion of a record of uncertainty in the risks?

Risk Analysis and Evaluation – To allow for a complete understanding of the risks, we need to identify, and question, the controls in place. This can be all be tabulated, but there are other tools that may allow better understanding and communication of the risks

Qualitative Bowties – Bowties are widely used in the process industries, and have been found to be very useful in several CCPS projects

Example Bowtie Structures – these are example approaches towards using bowties for understanding geological risks, the causes, consequences and typical (and type) of controls that might exist. Similar bowties have also been developed for other scenarios such as induced seismicity and releases from wells.

Example Populated Bowtie – these show examples of the information that can be shown on bowties.

Example Bowtie Extract Wells – bowties have also been used to look at the barriers in place to prevent leakage from wells, in this case the structure reflects the physical barriers in place for each identified leak path.

Semi-quantitative – It may also be useful to provide numerical estimates of risks e.g. to estimate mass released, to highlight highest contributing scenarios. A simple approach to adapt the LOPA method, to identify possible leakage scenarios and the geological barriers present (which may come from the HAZID and bowtie analysis). Estimating the likelihood that the sequence occurs and the probability of failure on demand of the barriers, allows for simple event trees to be constructed, which if combined with estimates of mass and duration of CO_2 release, allows for numerical risk estimates to be made.

Quantitative – wells – This approach has been applied to legacy wells of concern (e.g. high risk) and develops on from the leakage paths identified on the wells bowties. To provide a range of values, each physical barrier is estimated to be in one of three states, with a corresponding probability of being in that state and associated permeability. Solving Darcy's equation for each barrier allows for an estimation of a probability that the leak path exists and the mass of CO_2 that would be released.

Quantitative Methods Example Data – there is some data available for quantitative data, and these are some examples. The latter two represent summaries of several other studies, but still give a range of likelihoods and magnitudes.

Summary – existing tools can be adapted for use in CCUS risk assessments, but their limitations need to be understood.

142a Evaluating Energy Transitions to Achieve Net-Zero Goals

Pete Sharpe

The chemical industry's move to set aggressive Net Zero targets at the enterprise level has many sites scrambling to figure out how best to meet them. The easy, lower cost things like LED lighting, waste heat recovery, operational changes and Advanced Control come first, but at some point to get to Net-Zero, you need to start transitioning equipment to clean energy. For the chemical industry, that means (green) electrification of major combustion equipment, like compressors, furnaces and boilers, and moving away from fossil fuel generated energy systems. A common methodology that companies are using to evaluate and rank different "greening" options is called the Marginal Abatement Cost, or MAC Curves. This paper will present the MACC concepts and illustrate its use in several industrial examples.

142b SES Design: Towards Zero Carbon Process Utility Systems

Julia Jimenez Romero, Adisa Azapagic, Robin Smith

Rising energy costs and availability have accelerated the shift away from fossil fuels. In addition to reducing greenhouse gas emissions, the energy transition is essential for energy security and affordability. As one of the most energy-intensive sectors, the process industry is under significant pressure to enhance its sustainability. Although shifting to low-carbon source/technology may have positive and significant environmental effects, a framework for systematically analyzing the economics and practicality of the many options has not been thoroughly investigated. Along with viable solutions, energy transition requires a strategy to shift from existing systems to more sustainable ones over a time span bound by legislative and financial constraints.

This work presents an optimization-based tool for designing industrial carbon-neutral energy systems that can be constrained to reflect the design and operation changes over time. A multi-period optimization framework given a time horizon and a temporal resolution is built based on a proposed superstructure for carbon-neutral energy systems capable of minimizing the CO_2 emissions, whilst determining the optimal energy flows and associated costs. The bi-objective model is addressed by the ϵ-constraint approach, providing the decision-maker with a range of alternatives for developing a sustainable roadmap from existing systems to future ones.

The model identifies the best energy system configuration, seasonal operations, energy mix, and technology capabilities for meeting seasonal power, heat, and cooling demand in a sustainable manner. Based on a mixed-integer nonlinear programming (MINLP) model and life cycle assessment (LCA) principles, the model has been used to evaluate a wide range of technologies under different scenarios, including solar photovoltaics (PV), wind turbines, gas boilers, electrode boilers, gas turbines, carbon capture and storage (CCS) and energy storage.

The effective integration of conventional and renewable sources and the appropriate operation of energy storage units increase the system's flexibility and sustainability, while overcoming the intermittent nature of renewable energy sources. Emissions can be significantly reduced by 20%–40% based on current economics and restrictions. These numbers can be further reduced when electric power mixes present a higher penetration of low-carbon technologies. Overall, the framework can provide guidelines for economic and energy-efficient transition to low-carbon technologies and/or fuel switching in the industrial sector.

142c Three Differing Approaches to Industrial Decarbonization

Junius Jones

Abstract: Three Differing Approaches to Industrial Decarbonization

With the increased pressure to decrease CO_2 output, industry is looking at various approaches to reduce carbon intensity. Typically, decarbonization involves replacing traditional fossil fuel-based power generation with low-carbon or renewable power, such as wind, solar or biomass.

This presentation will focus on how the industry is innovating to meet the challenge of decarbonization from three different approaches to achieve a common goal of reducing greenhouse gas emissions.

We will explore the Origin Materials, PureCycle Technologies, and ION Clean Energy projects, which have taken different paths that utilize novel decarbonization processes at the forefront of their respective industries. Whether replacing products requiring petroleum-based feedstocks that emit CO_2 during manufacture with carbon-negative products, petroleum-based manufactured products with recycled products, or capturing emitted CO_2 from conventional processes, each company is poised to make a significant contribution to the reduction of industrial greenhouse gas emissions.

In addition to exploring these groundbreaking projects, we will discuss the key takeaways for delivering decarbonization projects successfully and the future of a decarbonized chemical industry.

Author Bio

Junius Jones is a Project Manager at Koch Modular and is responsible for design development and project execution.

Junius Jones has 25+ years of experience executing various sized projects from conceptual design to engineering, execution, and even operational services for many well-established companies in the chemical processing industry.

Prior to joining Koch Modular, he held various project engineering management roles at Fluor, Technip, Bechtel, and Worley. As a registered Professional Engineer in Texas, Junius holds a BS in Mechanical Engineering from Texas A & M University with MBA from University of St. Thomas.

151a Trade-Offs in Various Shipping Methods of CO_2 for Direction Injection to Wells in CCS Systems

Beverley Magumise, Ayuni Mohd Mohd Saidi, Arfahana Amir, Jon Barnes, Matt Healey, Eduardo Luna-Ortiz

Transport of captured anthropogenic CO_2 can be carried out in (offshore/onshore) pipelines or ships. The latter method is most efficient when the CO_2 needs to be exported over hundreds of kilometres or when pipeline infrastructure is not available/feasible. To date, most of CO_2 shipping operations are designed for port-to-port transportation or port-to-offshore with interim storage. Moreover, only refrigerated low pressure transportation has been considered.

In this work, we elucidate conceptual shipping schemes for direct offshore injection. We considered various vessel conditions including refrigerated transportation at low pressure (7 bar) as well high pressure (70 bar) and ambient temperature. We highlight the key challenges and limitations of various shipping regimes in terms of operating envelope, required fluid purity and strategy for unloading/injection in to the well.

151b Using Design Space Optimization of Software in the Loop to Minimize DRA Injection Opex and Maximize Pipeline Profitability.

Richard Merrett, Domenico Pinto, Arlie Nuetzel

Pipeline operators have a critical question to answer when it comes to addressing the changing energy market: How can OpEx be minimized in the transportation of oil. One way to achieve this is through the injection of drag reducing agents (DRA) to the lines. This can be used to either increase throughput or reduce pump load while ensuring product fluid properties are maintained. These chemicals are however expensive, and operating costs quickly mount up if the DRA injection isn't carefully optimized and controlled. Many operators currently use spreadsheets to try to understand how to optimize this, but this approach is time intensive and largely inaccurate.

In this presentation, a demonstration on how the optimization of DRA injection can be efficiently and accurately achieved with the use of Software-in-the-Loop (SiL) methodologies is given. With the use of a full fidelity, physics driven system model, in combination with design space optimization tools, the investigation of optimal throughput and DRA injection throughout a pipeline sample is performed. The tools enable the input of flow rates, energy and DRA costs to efficiently perform an accurate calculation of the overall cost per barrel of transport. The results of the investigation can then be deployed so, by using the optimal amount of DRA, any potential waste is drastically reduced which ultimately leads to increased profitability of the pipeline.

151c CO_2 Two-Phase Flow Operation in a CCS Transport Pipeline Network: Can We Trust Existing Multiphase Flow Simulators?

Eduardo Luna-Ortiz, Henry Tandoh, Kamila Szklarczyk-Marshall, Arsenij Fiodorov

The current wide-industry practice is that CO_2 transport via pipelines should be carried out in single-phase, either as gas, liquid, or supercritical state. The transport in any particular state has its pros and cons. However, the general rule-of-thumb is that is more efficient to transport CO_2 in liquid or supercritical form over gaseous phase. For optimal transport of CO_2 in supercritical/liquid phase, the minimum operating pressure should exceed the critical pressure for CO_2 mixture to avoid phase-splitting (two-phase flow). On the other hand, gas phase operation is an option at early life of the project when injecting into a depleted reservoir in order to minimize Joule-Thomson cooling. As pressure in the store increases, the system would then need to transition to high pressure operation to avoid entering into the two-phase region.

Conceptually, allowing two-phase flow operation during normal operation would offer certain advantages. It would potentially be the most cost-effective solution for the transition period and beyond (i.e. once gas phase operation is no longer possible). In this work, we present a conceptual study of an offshore transport system operating under two-phase flow regime for an industrial CCS hub located in North West England. The aim of the study is to try to answer two fundamental questions: Is the two-phase flow pipelines possible in a safe, manageable and predictable manner? Are there operational benefits in allowing pipelines to operate in a two- phase zone? In addition, we present a perspective on the current limitations of existing multiphase oil & gas flow models for CO_2 multiphase flow.

151d Conceptual Design of Profitable and Emission-Free Maritime Transportation

Ying Liu, Sujing Wang, Qiang Xu

Rising anthropogenic CO_2 emissions and global temperature have been great concerns in the world. Maritime shipping represents a large part of CO_2 emissions from anthropogenic sources. This paper has developed a novel conceptual design of profitable and emission-free maritime transportation with an inherent carbon capture process to produce the pipeline-grade CO_2 product. It employs liquid natural gas (LNG) and air as the major feeds, and retrofits and installs two industrial process subsystems in the ship including (i) an air separation unit (ASU), and (ii) an Allam-cycle power plant (ACPP). Through optimally integrating energy/material streams among two subsystems, this new design can simultaneously produce electricity to empower the ship, pure N_2, and pipeline-ready CO_2 products efficiently and effectively through eco-friendly manufacturing ways without CO_2 and NOx emissions. The newly proposed design has been modeled and examined through a commercial simulator, and its exciting economic and environmental performances have also been evaluated and demonstrated compared with traditional maritime transportation. The study has significant potential with salient economic, environmental, and societal benefits, which could be further optimized in the future in the maritime shipping field.

106a Rdh Eheater CO_2 Free Heating

Tuomas Ouni, Elina Nauha, Mikko Vermasvuori

Coolbrook is a Finnish-Dutch technology and engineering company on a mission to de-carbonize major industrial sectors such as petrochemicals and chemicals, iron and steel and cement. Coolbrook's revolutionary rotating technology combines rocket science, turbomachinery, and chemical engineering to replace burning of fossil fuels across all major industrial sectors. The technology has two main applications: Roto Dynamic Reactor (RDR) to reach 100% CO_2 free olefin production, and Roto Dynamic Heater (RDH) to provide carbon-free heating for process industry.

Rotodynamic Heater Technology: principle of operation, opportunities, and challenges

Coolbrook RDR and RDH share the same principle – they are electrically driven turbomachines where one or a cascade of stator-rotor-diffuser sequences heat gaseous feedstock efficiently into elevated temperatures. The heat increase occurs intrinsically in the heated gas and results from conversion of rotor blade -induced kinetic energy of the gas into thermal energy by rapidly slowing the velocity of the gas. As the velocity of the gas slows down from supersonic to subsonic, a shockwave occurs that enhances the conversion of kinetic energy into heat.

Coolbrook's RDH technology has potential to efficiently achieve process temperatures of up to around 1700 ºC for high volumetric gas flows and can thereby be used in processes that have been considered impossible to electrify. Main benefits of RDH are compact equipment size, competitive CAPEX and OPEX, and CO_2 emissions avoidance when powered by carbon neutral electricity. Additional advantages of RDH-based process heating arise from possibility to recycle the gaseous heating medium, >90% electricity-to-heat efficiency and ability of RDH to increase the gas pressure and thus operate as a combined heater/blower.

Pilot plant results

Coolbrook's Pilot plant in Brightlands Campus in Geleen, Netherlands is being built to test and develop RDR and RDH Technologies for electrified process heating and steam cracking of hydrocarbon feedstock.

Aerodynamic design of Coolbrook turbomachine has been done in cooperation with partners from academia and industry using state-of-the-art CFD modeling. Aerodynamics and rotor-stator-diffuser design dictate the performance of the machine in terms of heat and mass flows. Preliminary air testing of RDH/RDR power train was done in 2022 in Finland to experimentally verify the aerodynamic design and to demonstrate the principle of Rotodynamic Heating. At Brightlands pilot, the RDH test program consists of a variety of test runs with several gases, such as air, nitrogen, steam and methane, at temperatures up to 1000 °C.

Principles of aerodynamics design and preliminary results on the performance of the heater will be presented along with the general set-up of the pilot unit

Applications of Rotodynamic Heating in Heavy industry

A vast number of heavy industry processes require temperatures up to or over 1000 °C. Achieving such temperatures without giving rise to carbon dioxide emissions is a challenge that has not been properly addressed. Solution such as carbon capture, hydrogen firing or bio-based feedstock have proven costly in reducing the carbon footprint, and direct electrification of processes operating above 600 °C is challenging for electric heating technologies.

Rotodynamic heater provides a game-changing solution for compact and efficient electric heating of gaseous media into temperatures up to and above 1000 °C. Examples of the opportunities of Rotodynamic Heater in decarbonising the heavy industry include

- Hot gas generator for drying solid materials
- Heat provision for heat storages
- Revamp of existing furnaces for electric heat input

Future applications will include hydrogen heating for clean steel production, kiln heat generation for cement manufacturing, lime recovery or clay activation, and many other applications where hot gas from RDH is directly or indirectly used to heat high temperature processes.

106b High Efficiency Carbon Capture Solvent and Process for Low Pressure Flue Gas

Guofu Chen

High Efficiency Carbon Capture Solvent and Process for Low Pressure Flue Gas

The Challenge

Carbon Dioxide (CO2) is recognized as a major global warming contributor. Carbon capture and sequestration is part of the solution to reduce emission of CO2 to atmosphere. For high partial pressure of CO2, technologies are already very matured and it is a common practice for engineers to use MDEA and Piperazine to scrub CO2 from high pressure acid gas. However, for low pressure flue gas, such as gas turbine exhaust, we yet have not found a satisfying solution. A traditional approach is to use MEA as the solvent to absorb CO2 in the contactor, and strip off CO2 in the regenerator by heating. The heating duty of MEA solvent is about 4 GJ/tonne CO2. Do we have a better solution to reduce the heat load requirement to capture CO2?

The Solution

Engineers and scientists have been working on two different paths to answer the challenge. One path is to find the solvent that can absorb CO2 strongly and yet strip of CO2 easily with little amount of heat. The other path is to improve plant process design that integrates heat and power efficiently. This paper merges the two paths into one to deliver a complete engineering solution to capture CO2 from low pressure flue gas. The end result is much lower heating duty required to capture carbon dioxide, comparing to MEA.

106c Evaluation of CO_2 Capture on an Ethylene Plant

Myrian A Schenk

At Technip Energies we help our Clients to achieve their energy transition targets. We offer efficient technologies, project management, integration expertise and EPC capabilities which have been acquired over our long track record of projects in different industry sectors. Always, pioneering 'first of a kind projects', for example, progressing and delivering Prelude FLNG project and a fully modularized artic plant for Yamal LNG.

Similarly, for more than a decade, we have been working and delivering post-combustion carbon capture projects. Learning from that experience, is that the continuous improvement in the affordability of carbon capture is KEY to enable the industry expansion and for that, the integration between technology and engineering presents a huge step towards achieving this goal. We maximise the benefits of our alliance with Shell Cansolv® to provide a turnkey, single point delivery of projects. In this respect, we leverage in our more than 60 years of experience designing and building large capital projects across upstream, midstream and downstream industry traditional sectors (from power to Hydrogen, LNG, refining, petrochemicals) to the new industries, such as steel and cement that today also have the challenge to reduce their emissions.

We have built 'one team' Shell Cansolv® and TEN, that has enabled to deliver CapEx and OpEx reductions in our offerings of post combustion carbon capture plants. The team looks at the opportunities assembled with crossing the lessons learned in both organizations, and we aim to put together the best solutions for different markets and sectors. In the same way, throughout numerous projects we have developed common view on options of optimized heat integration between the carbon capture and the main plant, streamlining the impact on the process performance and overall guarantees. We do the same with regards to overall integration of facilities, for example waste water coming from the capture plant.

The main focus of the presentation will be on the details of adding a post-combustion carbon capture unit (CC) to an ethylene plant (EP). We will present our learnings through the years, considerations, concerns related to studies carried out for our ethylene clients. The idea behind adding a CC is that the EP should always work producing optimal amount of ethylene, while the CC will be designed to adjust and to adapt while the EP works at its best. We will discuss how the flue gas from the EP furnaces are connected to capture most of the CO_2 (up to 95% capture rate) and all the requirements of pre-treatment before entering the CC unit. We will also highlight how we propose to address the different modes of EP operation for the best operability. We will present the utility requirements and show ways of integrations with the EP. To conclude, we will present the technical cost of CO_2 capture and some show some alternatives CO_2 utilization or sequestration depending on plant location.

90a Solutions for Processing and Coprocessing Renewable Feedstock in Refinery Settings

Oussama Zenasni, Daniel Dreyer, Josep Marca, Maria Marquez

Drivers for the reduction of carbon dioxide production in the energy sector in the form of legislation in Europe (e.g., RED II) or economic incentives in the US and Canada, has led to an increase in conversion of renewable feedstocks to finished fuels and petrochemical feedstocks. Producers are utilizing both standalone biorefineries as well as existing fossil fuel plants to process these types of feedstocks, which include edible oils, waste oils (animal fats, used cooking oil, and crude Tall Oil), and pyrolysis oils. Compared to fossil fuels, renewable feedstocks present a new set of challenges, such as a higher risk for corrosion due to high TAN and chloride content, potential catalyst fouling due to impurities present in the feedstock, and variability in the properties of finished fuels (i.e., cold flow, lubricity, and conductivity). The effluent water from these units also poses new challenges in terms of oil/water separation and contaminants that need to be removed. NALCO Water solutions enable refiners to produce the desired products (finished fuels and petrochemical feedstocks) while mitigating the aforementioned risks. To that end, we have developed a line of offerings that addresses the unique challenges associated with processing and coprocessing a variety of renewable feedstocks and their corresponding finished fuels. Herein, we will discuss results associated with this product line and their application in the area of renewable fuels productions.

90b Theoretical Comparison of Energy Required to Produce Hydrogen By Different Technologies

Guofu Chen

Hydrogen is part of energy transition. There are many different technologies to produce hydrogen. Which technology is the best?

The answer depends on many factors, such as cost of material, cost of energy, safety, politics and so on. This paper strives to answer this question from a pure theoretical technical perspective. What's the minimum theoretical energy required to transform the feed materials into hydrogen?

Current technologies to produce hydrogen include:

- Electrolyzer
- Methane Pyrolysis
- Ammonia Decomposition
- Steam Methane Reformer
- Dry Methane Reformer
- Autothermal Reformer
- Partial Oxidation Reformer
- Coal Gasification

Exergy concept has been successfully implemented to analyze the maximum power a system can produce or the minimum power to change thermodynamic states. It has been primarily used by mechanical engineers when there is no chemical reaction. However, hydrogen production does involve chemical reaction. This paper will describe how to properly use exergy concept to determine the minimum power required to produce hydrogen. In the end, a comparison summary table will be presented on the minimum power required for each of the technologies.

90c A Scalable Process for the Production of Calcium Formate from Lime Mud

Omar Abdelaziz, Mats Galbe, Kaushik Chakrabarti, Christian Kugge, Hendrik Spod, Ola F. Wendt, Christian P. Hulteberg

The utilization of industrial side streams for generating value has always been of interest, but in today's society, with a high focus on the circular economy, it is becoming increasingly important. Lime mud (whose major component is calcium carbonate, $CaCO_3$) is a low-value by-product of the pulp and paper industry that is generated in considerable amounts during the causticizing process. Designing routes for lime mud valorization would present new opportunities for integrated biorefineries possessing an effective solid waste reduction. In the present study, a scalable process for the production of high-value calcium formate ($Ca(HCO_2)_2$) from $CaCO_3$ is developed. An alternative method is examined, where CO_2 and H_2 are used to react with $CaCO_3$ to be converted into $Ca(HCO_2)_2$ under mild operating conditions. The process uses a heterogeneous catalyst to react with the components in the system consisting of two solids, one liquid, and two gaseous compounds. The process development reported spans lime mud conditioning, catalyst development, process design and scale-up, including a preliminary techno-economic assessment.

AUTHOR INDEX

AIChE
120 Wall Street, FL 23
New York, NY 10005-4020

ISBN 978-1-7138-7938-1

9 781713 879381

CURRAN ASSOCIATES INC.
proceedings.com